# ARMORED TRAINS OF THE
# SOVIET UNION
# 1917-1945

*"Chernomoryets", an armored train of the Red Army which took part in the fighting for Tsaritsyn.*

# WILFRIED KOPENHAGEN

**Schiffer Military History**
Atglen, PA

# FOREWORD

Publications concerning armored trains are very rare. Official sources are as good as non-existent or inaccessible. When material concerning armored trains, a class of weapon not unjustly looked upon as dinosaurs in the field of military arms development, does appear, whether in the form of articles or larger features, there is generally a not inconsiderable response from interested readers. Many former members of this special arm surfaced following the appearance of German Armored Trains in World War II (Schiffer Publishing Ltd., 1989). Their help, together with new material, made it possible for author Wolfgang Sawodny to present almost all new photographs in German Armored Trains in World War II Vol.II (Schiffer Publishing Ltd., 1990). Since the German armored train force used captured Czechoslovakian, Polish, French and Soviet locomotives, cars and trolleys, the 112 photos and the drawings of the individual armored vehicles were of special interest beyond the German-speaking world. The publication was also a gold mine of information for modelers. The tables provided by Wolfgang Sawodny gave a concise overview of the disposition of the German armored trains and cars, the areas in which they operated and their whereabouts, as well as their armament and crews in the period 1939 to 1945. For this reason, therefore, and for reasons of space, this volume will not go into those Soviet armored trains, cars and armored cars on rails which were captured and used by the Germans.

Instead the author will concentrate on providing an overall view of the armored trains, armored trolleys and railway guns employed by the Soviet Union and their missions and areas of operation. The amount of material available made it necessary to limit the amount of text and photographs. It was therefore impossible to examine the designers and commanders of armored trains or the factories which produced the vehicles.

A description of the general history of armored trains and their common operating principles likewise had to be dispensed with in order to avoid a repetition of material already available. The following documentation is based solely on sources from the USSR. I am indebted to author Major Engineer Andrei Beskurnikov for his support in words and photos.

## PHOTO SOURCES

Author (4), Author's archive (48), "Trend" Archive (26), Hübner (2), Krzyzan (7).
Drawings: Modelist Konstruktor.
Title Painting: Heiner Rode.

## BIBLIOGRAPHY

Heigls Taschenbuch der Tanks, Teil III, Munich 1935.
V.A. Potseluyev: Bronenoszi shelesnich dorog, Moscow 1982
Group of Authors: Ognevoy metsch Leningrad, Leningrad 1977.
Group of Authors: Orushye Pobedy 1941-1945, Moscow 1960
N.A. Anoshchenko: Vozduchoplavatyeli, Moscow 1960.
W.W. Filipov: Vozduchoplavatyeli, Moscow 1989.
Group of Authors: Grazhdanskaya Woina i voyennaya Interventsiya v USSR, Moscow 1983.
Various volumes of the magazines Technika i Voorusheniye and Modelst Konstruktor (both Moscow).

Translated from the German by David Johnston.

Copyright © 1996 by Schiffer Publishing Ltd.

Printed in China.
ISBN: 0-88740-917-2

This book was originally published under the title,
Waffen Arsenal Sowjetische Panzerüge und Eisenbahngeschütze,
by Podzun-Pallas Verlag.

We are interested in hearing from authors with book ideas on related topics.

Published by Schiffer Publishing Ltd.
77 Lower Valley Road
Atglen, PA 19310
Please write for a free catalog.
This book may be purchased from the publisher.
Please include $2.95 postage.
Try your bookstore first.

*A Russian Army armored train of 1914.*

# INTRODUCTION

In very many cases history is downright curious. For example, historian Dimitri Volkoganov related that in 1921 Josef Stalin was highly indignant at then chairman of the Revolutionary War Council Trotsky, the supreme military commander. During his frequent trips to visit the fighting troops Trotsky not only surrounded himself with a large party of young, leather-clad Red Army soldiers, but also frequently had two armored trains accompany him. Stalin, who in any case envied Trotsky because of his talent as a speaker, his energy and his popularity, viewed this as a challenge. At that time he had no way of knowing that he – who could not stand flying –would make use of eight armored trains of the NKVD while travelling by rail to the Potsdam Conference in July 1945 (1923 km: USSR 1095 km, Poland 594 km, Germany 234 km). For completeness sake, it should

be added that there were six to fifteen sentries per kilometer of track. All told, the NKVD employed eleven regiments with 17,000 men as well as 1,515 men of its strategic personnel to guard the route. As to whether Stalin had a special personal relationship with these steel colossi on rails or whether the idea of guarding the route from Moscow to Potsdam originated from his entourage, the question remains open. As a matter of fact, one can say that this episode also marked the end of this class of weapon. The Soviets released little information concerning the command structures, organization, allocation of personnel and equipment, losses, total strength, the fate of the trains themselves or other details. After the war there were few clues as to whether plans existed in the USSR to continue using armored trains or other armored rail vehicles. Evidence of Soviet intentions in this direction was provided by the showing of the documentary film "Fruits of Victory" on East German television in

*Improvised armored trains were often equipped with a field gun as a forward weapon.*

*A 76.2-mm Model 1918 field cannon on an armored train on the Eastern Front.*

3

*Field gun on an armored, revolving platform.*

*Irkutsk 1918 — a locomotive fitted with armor plate by railway workers.*

**1975.** It showed a modern armored train: rolling down the tracks was an extremely squat train armed with electronically-controlled quadruple anti-aircraft guns and turrets from T-62 tanks, pulled by a diesel locomotive. Soviet tank officers offered the following personal opinions on this theme: the question as to whether armored trains were practical under current conditions was difficult to answer. At the time, however, it was the opinion of the officers that placing standard tanks on armored rail cars with heavy AA defenses would be conceivable, but that such a practice would only be used in order to reach a threatened sector more quickly when railroad

| | Length of Track (km) | | Increase Railway | Density |
| | 1900 | 1904 | (%) | km/100 km²) |
| --- | --- | --- | --- | --- |
| **Europe (excluding Russia)** | 282,878 | 305,407 | 7.6 | 3.0 (increase in USSR alone period 1922 to 1940 more than 22,000 km) |
| **Germany** | 51,391 | 55,564 | 8.1 | 10.3 |
| **Russia** | 48,460 | 54,708 | 12.9 | 0.9 (approximately 3,000 km in the western part of the USSR alone |
| **USA** | 311,094 | 344,172 | 10.6 | 4.0 |
| **Asia (excluding Russia)** | 60,301 | 77,206 | 28 | 0.3 |
| **Africa** | 20,114 | 26,074 | 29.6 | 0.08 |
| **Australia** | 24,014 | 27,052 | 12.6 | 0.3 |

represented the sole transportation link. That might be large expanses of desert- or taiga-like terrain. Subsequent to this now twenty-year-old documentary on postwar Soviet developments in the field of military technology, no further information on the theme of armored trains is known to have been released.

The origins of the armored train lie in the advent of railway troops and the railway net. Railway troops appeared in most European nations almost simultaneously with the construction of railroads. Although their designation, organization, strength and equipment varied from nation to nation, the missions of the railway troops were largely similar:

In peacetime they had to build strategically-important sections of track, improve their training methods in the process, eventually test new equipment for the railway industry, and erect artificial structures, including complicated bridges, more cheaply than the civilian forces. There follows a description of the railway net that existed at the beginning of the Twentieth Century.

In 1900 fourteen European nations had railway troops – smaller ones like Belgium, Bulgaria, the Netherlands and Denmark each had a company, Austria-Hungary had two battalions, Germany more than 27 companies and Russia a total of 35 companies each with four officers and 123 men (increased to 260 men in the event of war). The seven battalions of Russian railway troops committed in the Russo-Japanese War (1904-05) proved their ability there, even if the Russian Empire lost that military adventure. It is unconfirmed that the two Russian armored trains deployed then belonged to the railway troops, but it is to be assumed that they did. It is a fact, however, that they successfully covered the Russian troops against pursuing Japanese units and thus prevented even greater losses. It is worthy of mention that a Russian armored train had previously operated in the Far East. Along with 200 soldiers, it was a component of the Russian contingent of the international expeditionary force of major European powers which took Peking in August 1900 during the Boxer Rebellion (anti-foreigner uprising in northeast China 1900-1901).

Armored trains possessed no great significance in the Russian military until the First World War, although it is true that there were several on hand (the figures are contradictory: they vary between two and ten). During the war itself all participating states had armored trains (Belgium, France, England at least one each at the front, Germany 15, Austria 10, Russia 4); however on account of the rapid freezing of the front they scarcely played a role in the west. Armored trains were used primarily to guard the rear. As the personnel and weapons were needed at the front, the command in Germany and Austria reduced the complement of armored trains to 50%.

During the First World War Russian armored trains belonged to the railway troops and were relatively uniformly equipped: each of the Russian armored trains had two 76.2-mm guns and up to twenty heavy machine-guns. Initially the guns were not capable of all-round firing. They were installed on rotating pedestals on covered and armored flatcars and possessed only a limited field of fire to the side. The guns positioned at the front and end of the train could only fire in the direction of movement to the front or/and rear. This configuration subsequently came to be designated as an armored flat-car in the Russian Army. It was not until during the war were Russian armored trains fitted with rotating tur-

*This locomotive, part of Armored Train No. 49, was converted in 1919.*

*Armored Train No. 44, armed with 76.2-mm anti-aircraft guns, in March 1919.*

*Armored Train No. 85.*

*With its large-caliber cannon in revolving turrets, Armored Train No. 204 was the forerunner of the railway guns later produced in the USSR.*

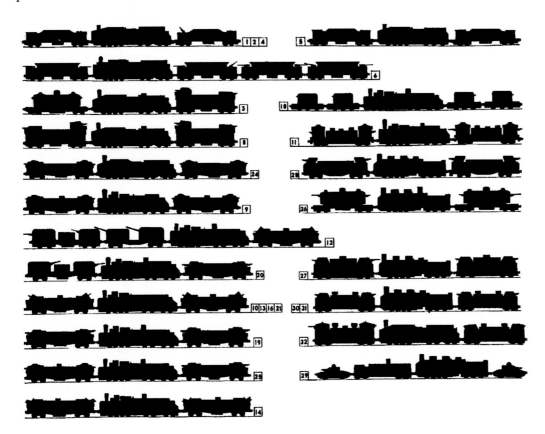

*Examples of the composition of Soviet armored trains:*

*1/2/4 —Anti-aircraft armored trains built by the Ishorsk Works (including No. 44 "Volodarskovo" and 3rd Petrograd "Mstityel"); 3 —No. 41; 5 —No. 67 "Bolgar"; 6 —2nd Petrograd; 8 —No. 87 "IIIrd Internationale"; 9 —No. 44; 10 — "Rosa Luxemburg"; 11 —No. 17 "Death or Victory"; 12 —No. 85; 13 — "Comrade Leo Trotsky"; 14 —No. 64 "Zentrobron"; 16 —No. 3 "Victory to the Soviets"; 18 —No. 4 "Communar"; 19 —No. 20; 20 —No. 71 "Akhtirets"; 21 —No. 7 "Stenka Rasin"; 24 —No. 45; 25 —No. 100 "Free Russia"; 26 —No. 98 — "Soviet Russia"; 27 —No. 27 "Burya"; 28 —No. 89; 29 —No. 96; 30 — "Karl Liebknecht"; 31 —"Communist"; 32 —No. 34A — "Red Soldier."*

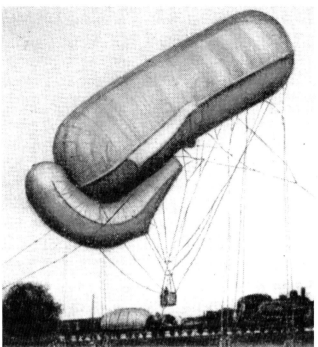

*A time-consuming operation — an observation balloon ascends from an armored train.*

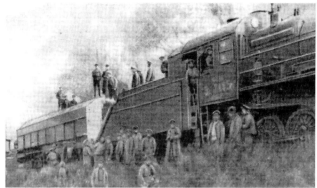

*May 1919 — the armored train "Volga" operated with the 23rd Balloon Battalion. Seen here is the armored balloon transporter*

*Top:  The 3rd Balloon Battalion was housed in Armored Train No. 9. The locomotive was not armored; however the railway car contained a balloon winch mounted on an automobile.*

rets with cannon for all-round fire. As the railway lines in the area near the front suffered great devastation in the early phase of the war, Russian armored trains saw little action at that time. According to Russian experts, contrary to the experience on the Western Front it was only when positional warfare came that the armored trains were able to support the ground troops effectively. Using shunting tracks along the front, they reacted quickly to changes in the situation and were able to hurry to threatened sectors of the front.

It was not until after the October Revolution of 1917, however, that armored trains saw really widespread use in Russia. This applied to both the White and Red forces and the intervention armies; and following their uprising, the captured Czechs (former soldiers of Austria-Hungary) fought from armored trains. Almost everywhere where there were large railroad workshops – in Moscow, Petrograd (now St. Petersburg again), Kharkov, Lugansk, Briansk, Tsarysino and Kiev – workers built "broneluchki", mostly for the Red Army. This was the designation for the initially makeshift armored trains built there. They consisted of coal lorries with cutouts in the sides for machine-guns as well as simple flatcars on which complete field guns were frequently installed. Railroad ties and sandbags formed the provisional armor; the locomotives were often left unprotected.

# THE STANDARD ARMOR TRAIN REMAINS THE SAME UNTIL THE SECOND WORLD WAR.

According to Soviet sources, the first action involving one of their armored trains took place on 20 November 1917, fifteen km north of Stanitsa Shlobin. It was there to suppress an uprising by a shock battalion of the White General Duchonin, who had attempted a mutiny in the high command of the headquarters. The armored train used its guns in support of an infantry regiment and a battalion of sailors. It is not known whether the later classic style of cooperation between the onboard shock troop and the train's own weapons with the infantry or cavalry developed there. In any case this combination, which was to remain standard practice in World War Two, and the arrangement of the weapons in rotating turrets or side bay-window gun mounts in order to achieve the maximum possible field of fire, developed in the course of the fighting in Russia. As well, the grouping of the cars around the locomotive, which was likewise gradually drawn into the armoring process, as well as the "controllers" at the front and back of the train – for safety in the event of detonating mines or as a defense against explosives-laden rail cars – were already standard at that time.

In the armored train the command of the revolutionary units, later designated the Red Army, saw a powerful weapon with which it could support its fighting units, which were poorly equipped in comparison to the troops of the interventionists and the White Guards. Moreover, the widespread use of armored trains was encouraged because most of the fighting in the initial period of the civil war was waged along the railway lines. There were almost no aircraft and the nation's road network was poorly developed. The railroad tracks thus provided the sole transportation link in the vast spaces of Russia.

One can say without exaggeration that armored trains took part successfully in all the major battles in Russia after the autumn of 1917. Frequently the workers not only built the armored trains but formed their crews as well.

In 1918 the Red Army had twenty-three armored trains; however by the end of 1919 this figure had risen to fifty-nine and toward the end of the war of intervention to 103. Several of these trains were repaired and modernized and saw service at the front in World War

*Soviet armored train from the 1930s: the narrow observation turrets between the 150-mm howitzers allowed the observers to see over the weapons turrets but scarcely limited the latter's field of azimuth.*

Two. As a rule the armored trains explored the situation as a sort of advance guard, engaged the enemy, covered the deployment of friendly infantry/cavalry into battle order, and supported these during attack or retreat with their fire. Armored trains possessed one major advantage compared to the field artillery: they could fire without loss of time and even while in motion and their fire was easily concentrated. They also carried with them a large supply of ammunition. Moreover, they offered protection to the gun crews and as well transported infantry units. They were thus capable of carrying out independent tactical missions. In several cases Soviet Russian armored trains even carried with them a tethered balloon with which to observe and direct fire. As a rule the locomotive was located in the middle of the armored train. Once the train had taken up position the commander was connected to all the battle sectors by telephone, enabling friendly cavalry and/or infantry to be supported effectively from the tracks.

Armored trains were first used in larger numbers during the fighting for Tsarytsyn (later Stalingrad, then Volgograd) in October 1918. With the help of a central fire control, it was possible to skillfully maneuver the twelve armored trains with their fifty guns so as to quickly go to the aid of the most threatened sectors. In an effort to improve the design of armored trains, after mid-1918 they were no longer built at the front but only in the workshops. Sheet steel six to eight millimeters thick replaced the railroad ties and sandbags. Armoring of

locomotives was now made universal, and all smaller caliber guns were installed in rotating turrets. Machine-guns were placed in side embrasures or in retractable bay window gun mounts so that they could fire along the train. However, several armored trains were equipped with 100-mm or 120-mm naval guns. These armored train batteries later served as independent units of the railway artillery within the framework of the coastal defense.

Armored trains were now produced mainly by the Putilov, Ishorski and Obukhovski works. In order to achieve a standardization of the existing diversity, on 31 January 1918 the military commission Zentrobron decreed that armored trains of the first class were to consist of a control car (which also carried tools, replacement parts and technical equipment) at the front and rear, a locomotive in the center, and a gun car (maximum caliber 75 mm) in front of and behind the locomotive. The arrangement of the cars of armored trains of the second class would be the same, but they were to be armed with weapons in the 100 to 150 mm class. Each armored train was to have a crew of 37 to 172 men and be armed with two to four cannon as well as four to sixteen machine-guns. In March 1919 the trains were merely classified as light or heavy, but after 1920 the classifications into armored trains type A, B and W took effect (see table).

## TABLE 1: SOVIET ARMORED TRAIN TYPES 1918 TO 1920

| Armored Train Type | | Personnel Complement | | | Armament | | Ammunition Capacity | | Components | |
| | | Total: | In Combat Units | In Service Units | Guns | Machine-Guns | Shells | Bullets | Armored Locomotive | Armored Flatcars |
| --- | --- | --- | --- | --- | --- | --- | --- | --- | --- | --- |
| Interim | 1918 | 95 | 71 | 24 | 2 | 12 | 400 | 72000 | 1 | 2 |
| | 1918 | 136 | 98 | 38 | 2 | 12 | 1350 | 216000 | 1 | 2 |
| | 1919 | 172 | 137 | 35 | 4 | 16 | 1200 | 210000 | 1 | 2 |
| Type "A" | 1920 | 162 | 137 | 25 | 4 | 16 | 1200 | 210000 | 1 | 2 |
| Mobile Armored Lorrey | 1919 | 15 | – | 15 | 1 | – | 80 | 1200 | 1(+) | 1 |
| Type "B" | 1920 | 15 | – | 15 | 1 | – | 80 | 1200 | 1(+) | 1 |
| Type "W" | 1920 | 37 | 21 | 6 | 1 | 2 | 160 | 6000 | 1 | 1 |

Command post in the control turret of a pre-1935 Soviet armored train. The position had means of communication to all important stations.

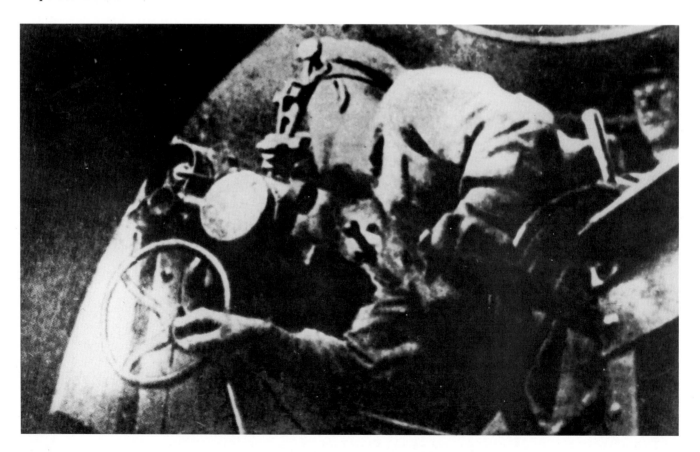

Gun crew in a 150-mm gun turret; the gunner is operating the elevating mechanism with his right hand while with his left he rotates the turret.

*An armored train, probably captured as war booty, in the 1930s.*

# DEVELOPMENTS IN THE 1920s AND 1930s

Following the introduction of a table of organization for the standard armored train, the depot in Nishni Novgorod and the bases in Tsarysino, Tambov, Koslov and Novosibirsk set about equipping such trains. These new units were used to form the reserve brigades.

The fundamental mission of the armored trains was seen by the military command to be the support of troops with artillery and machine-gun fire as well as the protection of railway stations and railway lines. The armored trains of the Red Army played a special role in the battle with Polish forces in 1920, especially with their mobile and extremely potent cavalry units. Alone, the Red infantry was helpless against them. The assessment of the Polish general Staff during the intervention of 1920: the Red armored trains are our most potent and formidable enemy, they are well-equipped, determined and brave.

After the experiences of the civil war and the intervention by foreign armies in Russia, the command of the Red Army saw armored trains as indispensable for engagements in the immediate vicinity of railway lines and rail junctions. As no large units with armored vehicles (tanks) were yet available, the red Army developed the following views regarding the role of their armored trains:

In an offensive role armored trains must operate in front of our own lines at the most threatened sectors with infantry units on board (normally two to three rifle platoons per armored train, about 90 soldiers with weapons) until our main forces arrive. If pursuing enemy elements, armored trains together with the cavalry must deprive them of any opportunity to regroup and dig in.

In a defensive role armored trains form a mobile fire reserve with which to strike the attacking enemy a mighty blow. They appear before him unexpectedly, destroy armored vehicles, artillery and infantry. During the retreat of friendly forces the armored trains provide fire support. Furthermore, they are responsible for protecting the railway lines and stations.

The armored trains belonging to the armored forces were grouped into battalions with two to three units and the battalions into armored train regiments. For communications duties these units were allocated machine-gun-equipped armored trolleys and later armored cars converted to drive on rails. before the war production of armored trains was limited to several types in order to achieve even greater standardization. Contracts for these were issued to several locomotive factories. Locomotives were required to be armored all-round and the production of anti-aircraft cars was called for as well as new artillery cars. In contrast to the fighting of 1917, powerful air forces had to be expected in any future conflict. In 1920 the installation of several machine-guns and occasionally cannon was adequate for anti-aircraft defense. The need for an integral anti-aircraft defense was not considered in the design of the BP-35, and on the Type NKPS-42 the anti-aircraft machine-guns were installed on the artillery turrets.

Table 2 provides a view of the composition and armament of the four main types of armored train from which the Type BP-43 was developed based on experience in the early phase of the war.

*Type BP-35 armored train (BP = bronepoyesd = armored train).*

*Armored rail car from the BP-35. Clearly visible are the "bay window" type gun mounts in the side walls of the car with Maxim machine-guns.*

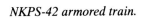

*NKPS-42 armored train.*

*Armored rail car of the NKPS-42.*

*A Red Army train carrying artillery weapons passes an OB-3 armored train at a station. On the locomotive tender is a four-barrelled Maxim machine-gun for anti-aircraft defense.*

*In this case the anti-aircraft weapon on the tender is absent.*

*Armored rail car from the OB-3.*

There were numerous modifications to the shape of the armor and turrets and the installation of weapons. Pictured here are three armored rail cars from the early phase of the war against the Soviet Union.

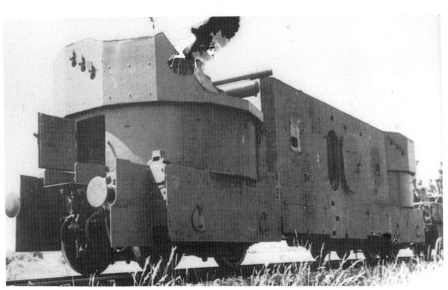

An official news agency photo of Soviet armored trains from the 1930s.

*Bottom Left:*

This armored trolley shows the marks of battle. Note the Maxim machine-gun for defense against low-flying aircraft.

*Bottom Right:*

German soldiers examine a captured Soviet armored train. As with others of its kind, it was almost certainly put into service by its new owners.

# TABLE 2: SOVIET ARMORED TRAIN TYPES 1941 TO 1945

| Armored Train Type | BP-35 | OB-3 | NKPS-42 | BP-43 |
|---|---|---|---|---|
| Armored Artillery Cars | 2 | 4 | 2 | 4 |
| Anti-Aircraft Flatcars | – | 2 | – | 2 |
| Security Cars | 4 | 4 | 4 | 4 |
| 76.2-mm Guns | 8 | 4 | 4 | 4 |
| 37-mm Anti-Aircraft Weapons | – | 2 | – | 4 |
| 7.62-mm Maxim Heavy Machine-Guns | 8 | 16 | – | – |
| 7.62-mm DT Heavy Machine-Guns | – | – | 12 | 12 |
| 12.7-mm DShk Anti-Aircraft Machine-Guns | – | – | 2 | 1 |

# PB-43 – ONE OF THE FINAL DESIGNS

The BP-43 armored train consisted of four security cars, four artillery flatcars, two cars equipped with anti-aircraft weapons and a PR-43 armored locomotive. The latter was a standard OW Series locomotive sheathed with armor plate. All mechanisms remained accessible via locking armored hatches. Observation sites and windows were also protected by armored hatches. Situated on the armored tender was the command turret as well as an open-topped position for a large-caliber DShK anti-aircraft machine-gun. The locomotive was equipped with a speaking tube as well as light and tone signalling systems. A radio set was installed in the command turret for external communication. The speaking tube link consisted of rubber tubes and metal pipes. It linked the commander of the armored train with the machinists, the commanders of the individual battle cars and the anti-aircraft gunners.

Mounted on each of the fully-armored artillery cars was the turret of a T-34/76 tank with an F-34 76.2-mm cannon and coaxial 7.62-mm DT machine-gun. Soviet designers opted to use complete tank turrets for the arming of armored trains and armored boats (this class of small warship existed only in the Red Army) for reasons of standardization. Based on the situation, the complete turrets of various types of tank had been used to arm armored trains. For the sake of completeness it must be pointed out that improvised armored trains of non-standard configuration were built during the Second World War just as they had been in the period after 1917. This

Up to 100 mm of armor in vital areas as protection against shells up to a caliber of 75 mm.
Maximum range with 10 tons of coal or 6 tons of diesel (in the case of a diesel locomotive) 120 km with a maximum speed of 45 kph.
Total weight of an armored train approximately 400 tons.
Armament
Range of travel of guns in turrets: 300 to 330 degrees horizontally, - 7 to + 40 degrees vertically.
Firing range of guns 8000 to 10000 meters, machine-guns 800 to 1000 meters.
Ammunition capacity: 76.2-mm guns – 280 shells 107-mm – 200 shells 37-mm anti-aircraft guns – up to 600 rounds 12.7-mm anti-aircraft machine-guns up to 10,000 rounds 7.62-mm Maxim heavy machine guns – 5000 rounds in belts

*Type BP-43 armored train equipped with standard T-34/76 turrets.*

even extended to armed street cars, one of which survives in the Odessa Museum.

However back to the BP-43 armored train: a DT heavy machine-gun in a ball mount was installed in each side wall of each artillery car. On top there was a hatch, and there was an armored passageway to the next car. There were two small double doors in the bottom section for exiting the flatcar in combat conditions. Observation was by means of direct vision slits in the walls and the vision devices of the T-34 turrets. Various loads, spare rails and ties could be transported on the free ends of the flatcars. Hatches accommodated entrenching equipment and tools for track repairs. Access from one car to another was via armored passageway.

Each anti-aircraft car had two non-rotating square turrets with armored sides and open at the top. In each turret was a Model 1939 37-mm anti-aircraft gun. While the train was on the move the guns were arrayed in the direction of travel with their barrels in the horizontal position. The upper part of the turret was protected by hinged armored shields which were folded down when engaging enemy aircraft. Each turret had three vision slits with armored shutters. Entry was from above over the side walls. Under combat conditions, however, it was possible to exit the turret by a two-part hatch in the bottom.

Materials for track repairs – rails, ties, connectors, bolts, etc. – were carried on the security cars at the front and rear of the train. The purpose of the security cars was to prevent the train itself from being damaged by exploding mines.

Each of the armored train's cars was steam heated by the locomotive – it was for this reason that the Wehrmacht gladly put into service captured armored trains made in the USSR.

*Above and below: Rail car from a BP-43 armored train. Legend: (1) armored roof (2) side wall (3) frontal armor (4) connecting passageway armor (5) spherical viewing port (6) observation slit (7-9) maintenance hatches (10) door (11-13) handrails (12) door bolt*

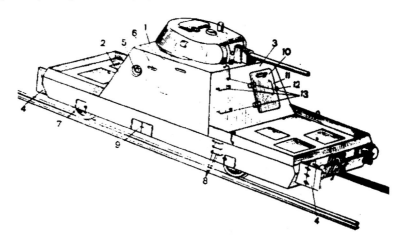

This armored train apparently fell into German hands intact. The side machine-guns and the anti-aircraft weapons on the tenders are pointed upward, and the hatches of the weapons turrets are open.

All the weapons along the entire length of the armored locomotive are also pointed upward.

*Complete turrets from the T-34/76 tank were not the only ones to arm Soviet armored trains; those of the T-26 or T-28 as well as other types were also used.*

*A photograph taken in March 1942 of a car from the armored train "Baltinets", armed with turrets taken from the KV-I heavy tank. Also conspicuous is the Maxim machine-gun in the car's front end.*

*"Michael", a captured Soviet armored train used by the Wehrmacht to patrol rail lines in the Crimea until May 1944. It is unclear whether the train had the mounted and faired T-34 turrets when it was captured or was later modified by the Germans. The concept of placing T-34s on flatcars and enclosing them was first employed by Moscow railway workers in 1942 (see also title page).*

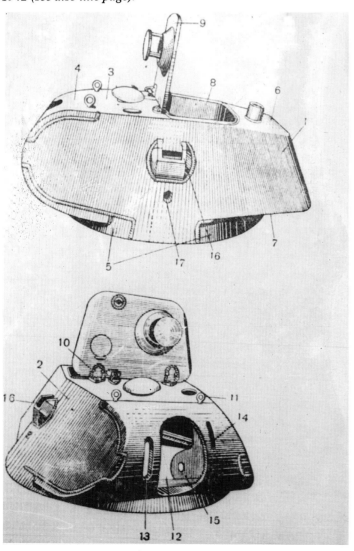

*T-34 turret as installed on the armored rail cars of the BP-43 armored train.*

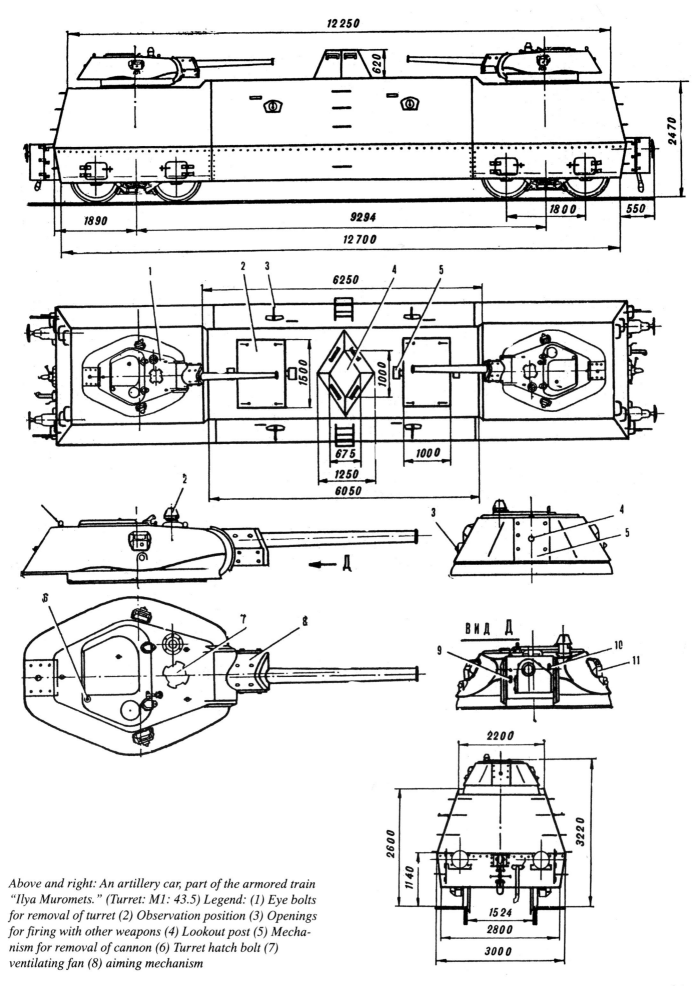

*Above and right: An artillery car, part of the armored train "Ilya Muromets." (Turret: M1: 43.5) Legend: (1) Eye bolts for removal of turret (2) Observation position (3) Openings for firing with other weapons (4) Lookout post (5) Mechanism for removal of cannon (6) Turret hatch bolt (7) ventilating fan (8) aiming mechanism*

The artillery cars of armored trains as well as self-propelled artillery cars were also capable of carrying infantry (below: a gun car captured by the Wehrmacht).

*Like the armored rail cars, the Soviet fleet of armored locomotives was extremely diverse: armored locomotives of the BP-43 . . .*

*. . . the OB-3 armored train . .*

*. . . and of the NKMS-42.*

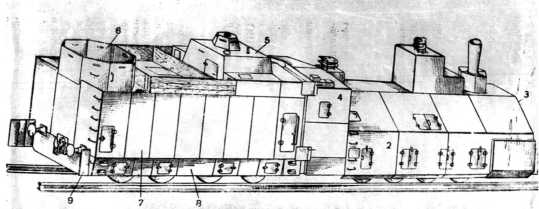

*Armored locomotive from the Type BR-43 armored train. Legend: (2) boiler side armor (3) front armor (4) fully-armored engineer's compartment (5) commander's command post (6) site for anti-aircraft weapons (7) tender armor (8) wheel cover panels (9) armor for passageway to next car.*

*The armored train "Krasnovostotchnik" on display in the Moscow Army Museum. It includes the OW Series locomotive No. 5067 built in 1896 and exhibits the typical layout of the BP-43 armored train. This locomotive was first armored by the Central Asian Railroad in 1917. It was subsequently modernized in September 1943.*

*Armored train armed with 76-mm guns. Its external shape suggests that it was originally Austrian equipment captured in World War I.*

*Captured by the Wehrmacht — a Soviet gun car armed with a single 107-mm field cannon. It was incorporated into Armored Train 28 in the fall of 1941.*

*Model of an armored train with an incompletely-armored locomotive on display in the Navy Museum in St. Petersburg.*

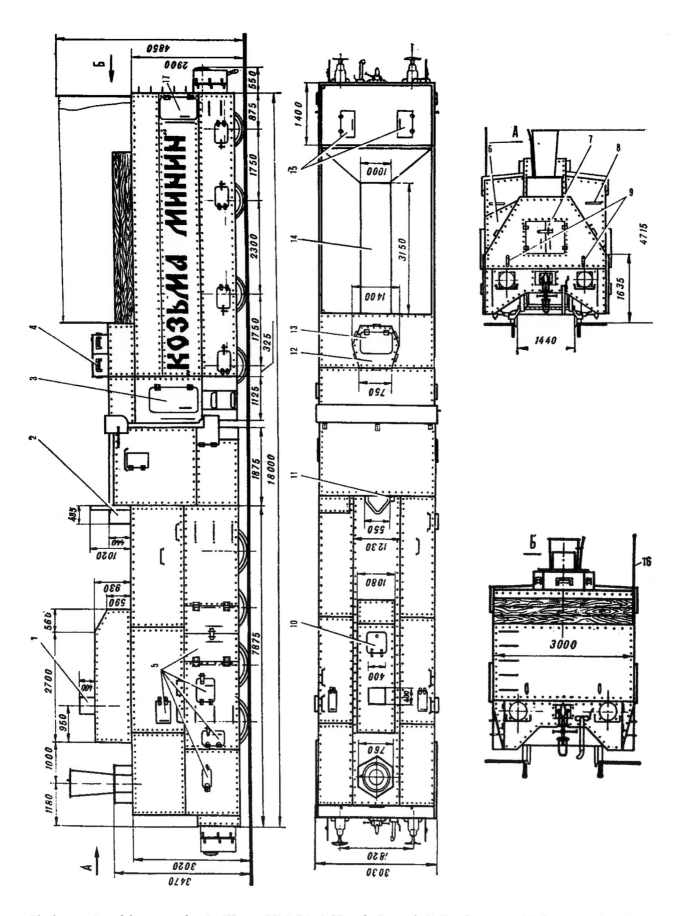

*The locomotive of the armored train "Kosma Minin" in 1:87 scale. Legend: (1-2) valve covers (3) door to engineer's compartment (4) optical observation devices (5) maintenance hatches (6) rod-linkage covers (7) door to boiler (8) vision slits for the engineers (9) searchlight mount (10) access to boiler (11) signal whistle cover (12) commander's turret (13) access hatch (14) coal access (15) hatches for taking on water (16)antenna (door to radio compartment.*

*This and top of following page: the Soviets fabricated numerous provisional armored trains and armored train locomotives after the outbreak of war.*

*Facing page: the installation of signalling and communications equipment was also diverse in nature.*

*Below: the commander of an armored train observes from the locomotive's command position.*

*A provisional armored train whose main armament consists of Maxim machine-guns; it was probably converted shortly after the outbreak of war.*

# POSTWAR ADVANTAGES AND DISADVANTAGES OF ARMORED TRAINS

After the war Soviet experts assessed the positive and negative aspects of their armored trains as follows (based on information provided the author by Major Beskurnikov).

Among the foremost advantages of this weapon was its speed of movement. On intact tracks, even facing threats from the air and ground, an armored train could cover approximately 500 kilometers in a day. Its armor was impervious to small arms fire and shrapnel. The train could therefore approach close to the enemy and engage him directly. At short range a minimum of at least four guns and eight to twelve machine-guns fired simultaneously from one side of the armored train. It also possessed its own means of anti-aircraft defense. In exceptional cases the armored train could also serve as a means of transport for the infantry.

The fact that the train was always tied to the railway lines was considered a disadvantage; as well even minor damage to the tracks could deprive it of its freedom of maneuver. Armored trains were more dependent on their support bases than other weapons. Water had to be taken on, coal had to be bunkered daily, and furthermore the boiler had to be cleaned.

Nevertheless – in the Soviet estimation – the armored trains of the USSR fulfilled their mission in the Second World War. According to Soviet authors, they played an especially outstanding role in the first phase of the war protecting rail junctions. Among those deserving of special mention were Armored Train No. 56 (barred the way to German tanks and motorized infantry which had broken through near Kiev), No. 72 (fought at Minsk, Briansk, Moscow, Leningrad and Stalingrad), "Ilya Muromets" (2,500 kilometer battle route as far as Frankfurt on the Oder; seven aircraft shot down, one armored train and seven mortar/gun batteries destroyed), and "Uzbekistan" (reached Brandenburg at the end of the war).

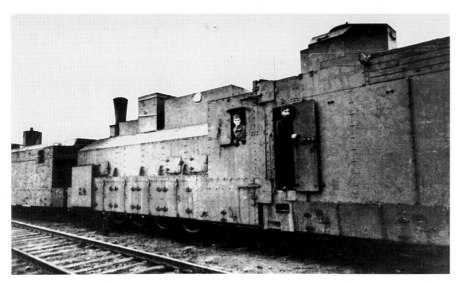

*Left: As well as undamaged armored rail cars, the Wehrmacht added intact locomotives to its fleet.*

*Even at the beginning of the war this locomotive with tender was armed with an anti-aircraft gun (probably 25 mm caliber).*

# SPECIAL ANTI-AIRCRAFT ARMORED TRAINS

Although the Soviet side stressed the effectiveness of armored trains as concentrated and very maneuverable sources of firepower in the initial period of the war – roughly until the end of 1941, early 1942 – their losses were not to be overlooked. There are no statements available as to whether mines, tanks and artillery, or aircraft played the leading role in putting a number of Soviet armored trains out of action. Not only did tanks demonstrate their superiority over the armored train on the battlefield; as a result of rapid development during the war their armor protection and the caliber of their guns surpassed the protection and armament of the armored

trains. In spite of this conflicting situation, given the difficult situation facing the Soviet armed forces it is quite understandable that they did not remove their armored trains completely from service, but instead used them to bolster the firepower of the ground troops at suitable locations. Anti-aircraft armored trains played a significant role in the course of subsequent battles. Their composition, their armament and their missions may have been predicated on the concrete demands of the sectors of the front in which they were created and/or employed. Based on available photographs the anti-aircraft component of regular armored trains was bolstered through

*Improvised self-propelled anti-aircraft guns: a GAZ-AA truck with a four-barrelled Maxim mounted on a flatcar could be coupled to armored trains to provide anti-aircraft defense.*

*The Soviets obviously soon recognized the role of the air observer.*

*Above and left: The 7.62-mm Maxim ma-chine-gun was the armored trains' main weapon against low-flying aircraft in the early phase of the war.*

*Above: Soon, however, the Soviets added 12.7-mm machine-guns and larger caliber anti-aircraft weap-ons. Special anti-aircraft trains armed with several 85-mm anti-aircraft guns were used to protect special installations.*

*Four-barrelled Maxim guns in firing position. When not required, the anti-aircraft weapons could be covered by hatches (center of photo).*

the installation of four-barrelled Maxim machine-guns. Pure anti-aircraft armored trains on the other hand were armed with anti-aircraft weapons of 12.7-mm caliber and greater. There, too, it is certain that the material on hand was used.

Armored trains of this type were used in coastal areas and near ports – often manned by sailors – as well as in the interior of the country.

Soviet sources stress their special role in covering railway junctions, bridges and other very important rail installations.

In a revue which appeared in its 9 January 1986 issue, the army newspaper Krasnaya Svedsa (Red Star) reported that the independent anti-aircraft armored train Bolshevik had covered a total of 13,000 kilometers in the period from January 1942 until the end of the war and shot down 30 German aircraft. According to the article the armored train had light and medium cannon as well as super-heavy machine-guns for anti-aircraft defense. The officers commanding the individual battle sectors were trained anti-aircraft artillerymen. The train was subordinate to the respective division sectors of the territorial air defense. In February 1942, for example, it was under the command of the Voronezh – Borisoglebsk Air Defense Division. Women made up part of the crew, being employed as gunners and telephone operators.

As in regular anti-aircraft units, the armored train included a command post, air observers, and range-finder and communications platoon, as well as the flak batteries, which were divided into fire platoons.

Also worthy of note is the unique weapons array of tank turrets (T-34), anti-aircraft guns and rocket launchers (Katyushas to the Russians, "Stalin Organs" to the German soldier) carried by several Soviet armored trains.

One such was the armored train Kosma Minin, built in Murom from plans by the Gorki Waggon Works in the period October 1941 to February 1942. It is possible that the M-8 launcher for the Katyusha rockets was installed at a later date.

*Anti-aircraft car with two Model 39 37-mm cannon in simulated firing position. The upper armored panels would have to be folded down if the guns were in fact to be used.*

33

*An armored train of the Navy in the Oranienburg area (March 1942) with weapons in the "anti-aircraft defense" position. Like the small-caliber cannon behind it —almost certainly taken from a warship —the 12.7-mm weapon in the foreground has an armored shield for use in the ground role.*

*Karelian Front, May 1943: made up of improvised cars of several different types, this anti-aircraft armored train has taken up position. Armament consists of DShK anti-aircraft machine-guns (12.7 mm).*

34

*Armed with two Model 39 37-mm cannon, this anti-aircraft car was part of the standard equipment of the Type BP-34 armored train. There were, however, variations in the height of the armor and the side protection provided the chassis.*

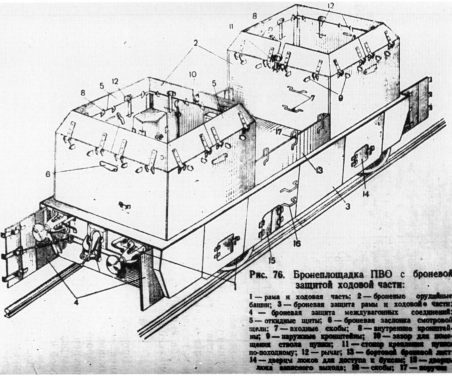

Рис. 76. Бронеплощадка ПВО с броневой защитой ходовой части:

1 — рама и ходовая часть; 2 — броневые орудийные башни; 3 — броневая защита рамы и ходовой части; 4 — броневая защита междувагонных соединений; 5 — откидные щиты; 6 — броневая заслонка смотровой щели; 7 — входные скобы; 8 — внутренние кронштейны; 9 — наружные кронштейны; 10 — зазор для помещения ствола пушки; 11 — стопор крепления пушки по-походному; 12 — рычаг; 13 — бортовой броневой лист; 14 — дверцы люков для доступа к буксам; 15 — дверцы люка запасного выхода; 16 — скобы; 17 — поручни.

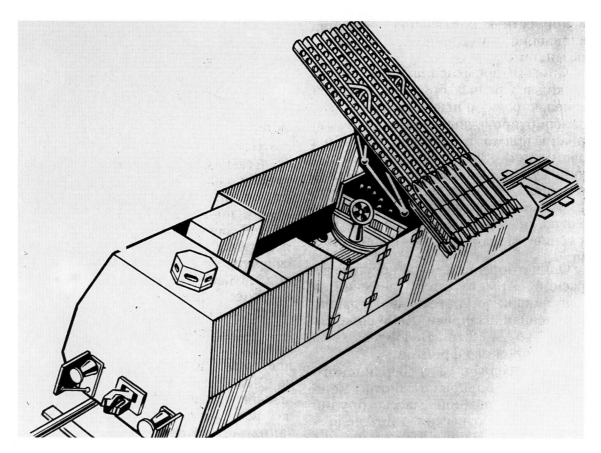

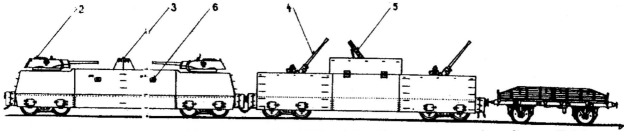

According to official statements by the Soviet trade press several armored trains were also armed with rocket launchers —the so-called "Katyushas" or "Stalin Organs." The model photo illustrates the rocket launcher installation and the line drawing the combination of rocket launcher and anti-aircraft weapons on a car forming part of the armored train "Kosma Munin."

A title illustration from the magazine "Modelist Konstruktor" (Moscow, May 1980, original in color) depicting the armored train "Kosma Minin." The latter was remarkable if for no other reason than its M-8 launcher for 82-mm rockets.

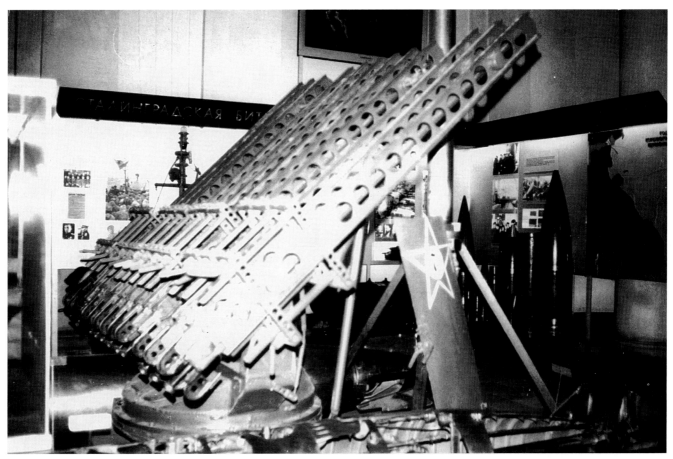

*The rocket launcher — here on display in the Navy Museum in St. Petersburg — was also mounted on warships.*

*Below: 1:87 scale drawing of the combined car for use by modelers. (1) M-8 launcher for 82-mm rockets (2) folding armor panels (3) vision slit (4) 37-mm anti-aircraft gun (5) hatches in cabin for M-8 crew.*

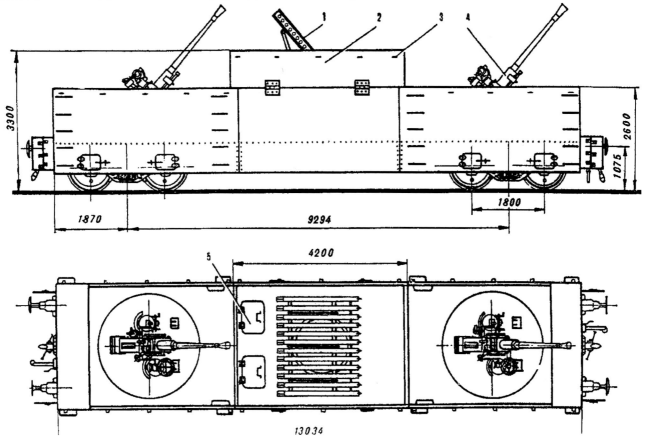

# RAILWAY GUNS – STANDARD AND IMPROVISED

Several of the Soviet Russian armored trains on strength after 1917 were more aptly to be classified as railway guns, however little importance was attached to this at the time. One example of this is the combination from the year 1920 which was designated as the armored train Ataman Churkin; it consisted of a four-axled car with a forward-facing large-caliber, long-barrelled cannon in front of the locomotive and several passenger cars. Whether armored and armed cars also belonged to the train cannot be determined from the photograph. One innovation is discernible, however: the train is carrying a captive balloon from which an observer corrected the gunfire and observed surrounding terrain within a radius of roughly twenty kilometers. A number of other Soviet armored trains also had such "ears" and "eyes" at that time. The balloon was released from an armored rail car, whose interior contained the winch.

From the mid-1920s Soviet designers devoted themselves to the development of new railway guns. In 1927 engineer A.G. Dukelski began work aimed at placing the well-known 356-mm cannon from the cruiser Ismail on rails. At Dukelski's suggestion a special bureau for the design of railway guns was established in 1930. By that time the nation's economy had become stronger, and starting in 1924/25 the Soviet military began a general program of reform and modernization. Based on existing naval artillery, the USSR produced railway guns with calibers of 130 mm (range 23.5 km), 152 mm (range 30.8 km), 180 mm (range 37.8 km) and 356 mm (range 31.2 km). The aim of this development program was to guard the nation's lengthy borders together with the existing or yet to be built forts and coastal artillery batteries. Since the railway guns had to operate together with the ships and boats of the naval forces, they too were under the command of the navy. And so first the Baltic Fleet and then the Far East Fleet were equipped with railway guns. An extensive track system along the coast with many side-tracks and dummy positions was planned in order to give the rolling artillery optimal maneuvering potential. There is no information available as to what extent this plan was actually put in place prior to the outbreak of the Second World War. In any case the Soviets obviously did not form and equip the planned number of batteries.

The USSR's first new railway gun was a 356-mm cannon. The vehicle had a mass of 340 tons. In 1932 the first two batteries, each consisting of three guns, were stationed on the Pacific Ocean. A short time afterward work began on a 180-mm railway cannon which was to be capable of engaging enemy warships at great range. In 1933 two railway guns with a caliber of 305 mm entered service. At the start of the war the Soviet coastal defense had available eleven batteries with thirty-seven railway guns (6 x 356 mm, 9 x 305 mm, 20 x 180 mm).

It should be mentioned that the Soviet military conceded a solid place to the railway gun. At the same time the designers also worked to equip the field artillery with large-caliber weapons while retaining its mobility. For example, they produced the 203.2-mm howitzer which rested on a tracked carriage and was to be towed by a

*A naval gun on rails during the revolution and civil war; it was the predecessor of later Soviet railway artillery.*

*Facing page: Two different artillery cars from the armored train "Mstitel" from the same period in 1:87 scale.*

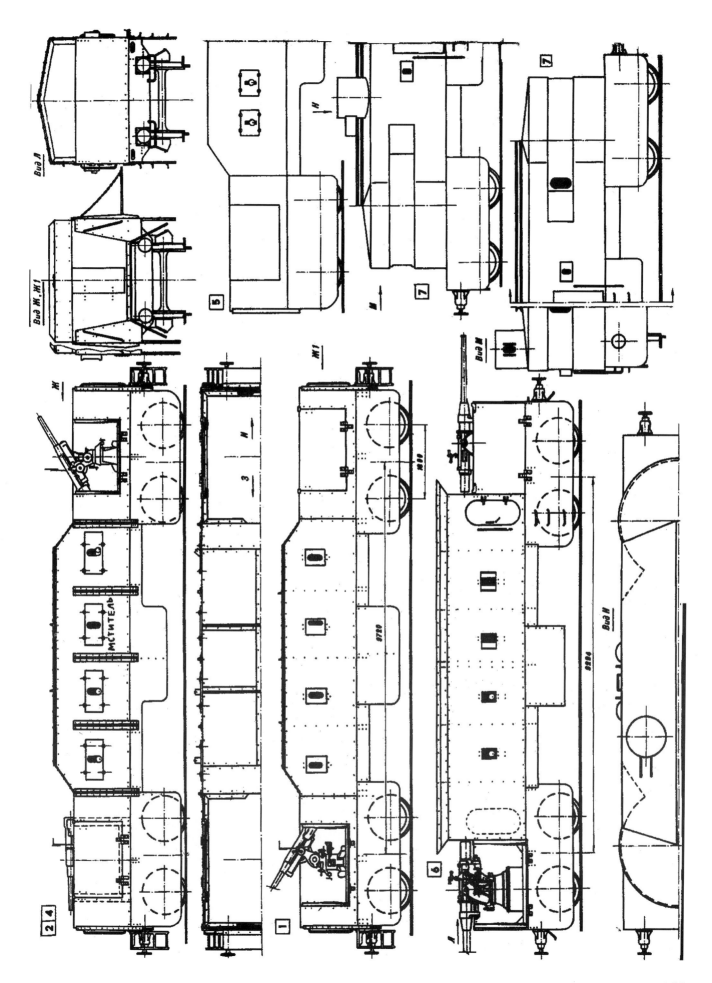

There was no comparable weapon in any other country.

As planned, after the German invasion the coastal and railway batteries operated together with the naval artillery to support the land forces or one other in large operations. One such example was the defense of Leningrad.

The ships' guns, the coastal artillery, and the cannon mounted on railway cars began to deploy when the German forces neared the city on the Neva. Initially this included the four heavy railway batteries of the Baltic Fleet which were included in the defense of Leningrad. As well there were the twenty-nine batteries with seventy railway guns which had been built at short notice after August 1941: all of the 100-, 130- and 152-mm guns in the fleet's artillery depot and those planned for new ship construction were mounted on suitable railway cars and made capable of firing. This was not without precedent – an examination of photographs from the period up to the beginning of the 1920s reveals that they had also placed complete naval guns on the railroad at that time. Added to this already extensive arsenal were the weapons from damaged or obsolete warships, even including those of that sacred relic of the revolution the cruiser Aurora. Among them were numerous 120- and 180-mm cannon. In the case of single-gun turrets, these were installed by placing the complete turret on a railway car. The majority of Soviet railway guns operated around Leningrad, taking part in the fighting together with two armored trains (from a total of six which had belonged to the coastal defense until the war) and the guns of the fleet units. The effectiveness of the heavy shells and the firing range of the guns on rails exceeded those of the field artillery. Although the track system was very often destroyed and a great deal of time had to be expended camouflaging the cars, the batteries on rails constituted a very effective firepower for the defense of the city. The dense system of tracks in and around Leningrad and newly-constructed artillery positions were exploited to move the guns unexpectedly into firing position, place surprise fire on the enemy rear, headquarters, reserves, fortifications and troop concentrations or lay down barrage fire. Formed on 8 January 1942, the 101st Naval Brigade of the Naval Artillery placed the individual batteries under a unified command. This unit was renamed and reorganized several times as the situation required. That same year the sixty-four 130-, 152-, 180- and 356-mm railway guns were concentrated into battalions and each was placed under the command of a unit of the land forces. According to Soviet sources, in the Leningrad area in the period 1942/43 the brigade expended 144,000 shells; the majority of these were aimed at distant targets, but some also served to cover the sea route between Leningrad and Kronstadt and suppress the enemy's artillery.

Available for fire correction were a special airforce squadron and the 3rd Airship Battalion with several tethered balloons from which observers spotted important targets behind the front lines (supply routes, transport

*This railway gun has been preserved in Dniepropetrovsk as a technical monument. It was placed into service in the city in 1918 and was used in the fighting in the Ukraine.*

*A true railway gun (caliber 356 mm) during the Second World War in the Leningrad area in firing position.*

*Complete gun turrets from warships, like this 130-mm weapon, were placed on railway cars.*

*130-mm naval gun on land in firing position. Very little of the rail car may be seen.*

installations, troop concentrations).

By October 1943 the Soviet command was in a position to withdraw some of the navy railway artillery from Leningrad. The guns were used to reinforce the coastal artillery, returning to their true intended role. Individual railway guns had already fought in other sectors of the front. The 200th Battalion saw action in front of Moscow with twenty 100-, 130- and 152-mm railway guns as did the 193rd Battalion with ten 100- and 152-mm guns.

An independent 152-mm battery operated near Stalingrad, and railway guns also intervened elsewhere in support of the ground forces. For the rest of the war railway guns from the Leningrad group joined the advancing Soviet forces, where their role was the destruction of special targets. In 1944 they helped reduce the fortifications of Vyborg and silenced artillery positions near Memel (present day Klaipeda). Two railway artillery groups (one with ten 180-mm, nine 130-mm and twelve 152-mm guns, the second with seventeen 130-mm guns) subsequently operated near the coast. In April 1945 several batteries took part in the battle for the cities of Königsberg and Pillau, bombarding the forts which lay beyond the range of the field artillery. From December 1943 until May 1945 a total of six battalions with sixty-two railway guns with calibers of 130 to 180 mm fought with the Soviet land forces. They expended 15,028 shells in 837 fire missions, in the course of which they sank eight ships and damaged five, destroyed seven transport trains, and reduced the fortifications of twenty-two strongpoints (all dates, numbers and statements from Soviet sources).

Little was heard of Soviet railway guns after the war. several generations of coastal defense rockets took their place, and all that was to be found of the guns on rails was a few photos and models in the country's naval museum.

However, at least one railway gun has survived intact and is accessible to the public. It is a 152-mm cannon with revolving carriage and turret, which rests on a siding of the Sevastopol railway station.

*Model of a railway gun in the former Soviet Navy Museum in Tallin.*

*Top of facing page: This 305-mm railway gun was designated as the TM-2-12 Transporter.*

*Bottom of facing page: 180-mm railway gun in Leningrad.*

# TECHNICAL DATA OF SEVERAL SOVIET RAILWAY GUNS

In Russian parlance a railway gun is a transporter.

**Transporter TM-2-12**

| | |
|---|---|
| Caliber | 305 mm |
| Barrel length in calibers | 52 |
| Barrel elevation angle | 30 deg (45 deg) |
| Gun weight | 470.5 kg |
| Muzzle velocity | 823 m/sec |
| Firing range | 26.5 (30.2) km |
| Length of transporter | 33.72 m |

Data in brackets: following modernization of the Model 1910 gun.

**130-mm Gun of the Coastal Artillery**

| | |
|---|---|
| Weight | 12.8 t |
| Projectile weight | 34 kg |
| Firing Range | 25.6 km |
| Rate of Fire | 7-8 shots/min. |
| Crew | 11 men |

(was the main armament of Soviet destroyers built in the 1930s and 1940s: Gnevny, Storoshevoy and Leningrad classes).

# ARMORED CARS ON RAILS AND ARMORED MOTOR CARRIAGES

The extensive experiences with armored trains after 1917 had demonstrated the need for small, extremely mobile and, if possible, also armored rail vehicles for various communications and supply missions, but also for reconnaissance. The result was the use of a wide variety of trolleys. The Soviets did not limit their efforts to that, however. Experiments were undertaken to equip available automobiles to ride the rails on their rims or remove the wheels completely. Armed with a cannon and three machine-guns, the Garford-Putilov armored automobile was used on rails during the civil war. When, in the course of the reorganization of the Red Army's railway guns, plans were made for armored reconnaissance and communications vehicles for these units, the Soviets drew on this experience and used the BA 20 light armored automobile and the BA 6 and 10 heavy armored automobiles with interchangeable wheels for road and rail travel. The Wehrmacht captured large numbers of these vehicles in 1941 and put them to use itself.

As rearranging the interchangeable wheels which were always carried externally by the BA 20 Shd (Shd – Russian abbreviation for railway), BA 20 ShdPU (with external antenna for radio set), and BA 6/10 armored automobiles proved very complicated and time consuming, experiments were also carried out with front and rear folding axles which were used only on rails. The normal pairs of wheels were retained; these also ran on the rails and propelled the vehicle. Experiments of this nature were carried out in 1943 with the BA 64 armored automobile (built on the GAZ-67B jeep chassis); however the BA 64 Shd did not enter production. The principle of placing standard vehicles on rails with the help of small guide rollers on falling axles was retained after 1945 in the Soviet Union, especially by the railway pioneers, and was used on a large scale. In particular, the vehicles required by railroad bridges were equipped in this way. Prior to the Second World War tactics for the employment in combat of Soviet armored trains envisioned the use of these armored automobiles to reconnoiter ahead of the train to a distance of ten to fifteen kilometers. Two BA-20Shd and three BA-6Shd or BA-10Shd armored automobiles were assigned to each armored train. Experience with these armored automobiles on rails may have led to the desire to create a combination of armored train and armored automobile. On the one hand a requirement existed for small armored

*The Garford-Putilov armored automobile was used on rails as an armored trolley.*

*Based on the GAZ-AAA truck: the BA-6 armored automobile, here the BA-6Shd version converted for use on rails.*

and armed rail vehicles, on the other they were to have greater operational capabilities and combat value than the armored automobiles but not the size of a complete armored train. At the end of the 1930s the Kirov Works in Leningrad designed and built the motorized armored car (MBW - Motorniy bronyevoy wagon). The designers used several components of the T-28 medium tank: in the areas of armament and the drive train. The following external features were characteristic of the MBW: there were three pairs of wheels beneath the front end of the vehicle (or to put it another way, the end facing in the main direction of travel; obviously it could also travel in the other direction). These were necessary to support the weight of the two complete turrets from the T-28 mounted in a stepped arrangement. The rear machine-gun was omitted from the lower turret. Otherwise all three turrets, including the one behind the command turret, retained all the features of the tank turret, including the searchlight on the barrel of the PS-3 cannon and the coaxial DT heavy machine-gun. The extent of traverse of turrets 1, 2 and 3 was 280, 318 and 276 degrees respectively. The guns' vertical range of movement was -5 to +25 degrees. In addition to the total of five DT machine-guns, two Maxim guns were placed in flexible mounts in each side wall. The DT heavy machine-guns in the rear of turrets 2 and 3 had a field or traverse of 30

to 17 degrees and a vertical range of -40 to +50 degrees. The forty-man crew had at its disposal 365 76.2-mm artillery shells, 145 magazines for the DTs with 10,962 rounds and for the Maxim machine-guns 48 belts with 250 rounds each and 20 belts with 500 rounds (a total of 22,000 rounds). Available for fire control were the PT-1 Model 1932 tank periscopes and TOD Model 1930 telescopic sights mounted in the turrets. The latter could be rotated electrically or by hand. The range-finder optics possessed a magnification factor of four.

The entire vehicle was armored. The armor of the side panels, which were sloped 10 degrees at the top, was 16 to 20 mm thick. The roof was 20 mm thick, the hatches 10 mm, and the turrets themselves 20 mm. The MBW was commanded from the center turret, on which the AA Maxim could also be posted. Entry was by way of three doors per side as well as the turret hatches. Turrets 2 and 3 were also able to accept one machine gun each (DT heavy MG) for anti-aircraft defense.

The M17-T gasoline engine (a tank motor derived from an aircraft engine) had an output of 294 kW (400 HP). It gave the 80-ton MBW a maximum speed of 120 kph. The vehicle could pull loads weighing up to 120 tons. The most important elements of the drive train of the T-28 tank were incorporated into the design of the MBW. These assemblies were installed in the rear part of the

45

*BA-20Shd armored automobile used by the Soviets as a light armored trolley, seen after its capture by the Wehrmacht.*

vehicle. Electrical equipment consisted of two GT-1000 and one PN-28.5 generator as well as eight 6STE-128 storage batteries. External communications were provided by a 71-TK-1 radio station, while internally there were six telephone stations. In addition, there was a powerful light signal system. At the start of the war the MBW, which was produced in limited numbers, was assigned to the armored train battalions, which were under the Tank and Mechanized Troops Command. According to Soviet sources several vehicles were equipped with T-34 turrets as well as the tank's W-2 diesel engine. One MBW has been preserved as a technical monument; it has T-28 turrets but is equipped with the 76.2-mm cannon of the T-34. There is no information available as to whether this armament combination was installed during construction or was the result of a subsequent conversion.

Other factories in the USSR built armored rail cars similar to the MBW. There was one type with a 180 HP motor, two cannon and four machine-guns (range 500 km, max. speed 60 kph, crew 21 men, weight 34 tons, armor 20 mm). Six such vehicles were captured and put into service by the Wehrmacht.

For the sake of completeness, it should be mentioned that the turrets of other tank types were used to arm armored rail vehicles according to the situation. There are photographs of Soviet armored trains armed with the turret of the T-26 tank and its 45-mm cannon.

In addition to museum models of the rail-bound weapons described in this volume and the railway gun in Sevastopol, there is an armored train in the Moscow Army Museum assembled from components from several units which gives a vivid impression of these giants of the technology of the past.

## WEAPON TYPE

| Cannon | PS-3 Model 1927/32 | F-34 Model 1910 |
|---|---|---|
| Caliber | 76.2 mm | 76.2 mm |
| Barrel Length | 16.5 calibers | 41.5 calibers |
| Muzzle Velocity | 381 m/sec | 662 m/sec |
| Weight of armor-piercing shell | 6.5 kg | 6.3kg |

*Inspired by the Soviet armored trolleys — armored trolley built by German troops with the turret from a Soviet light tank (caliber of gun 45 mm).*

*Converted for use on rails and tested, but not produced in quantity — the BA-64Shd armored automobile.*

*The crew of this armored self-propelled rail car obscure the two gun turrets which were step-mounted one behind the other.*

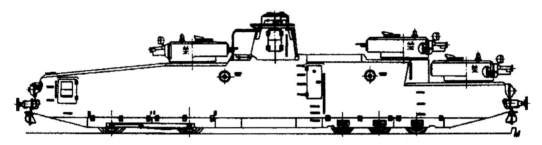

*Three complete T-28 turrets formed the main armament of this "motorized armored rail car."*

*The turrets of this version were equipped with T-34 cannon.*